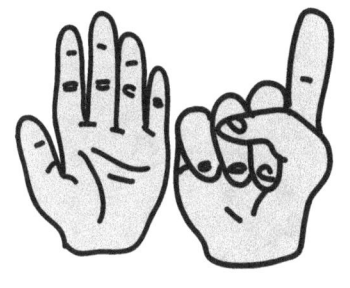 Six 6

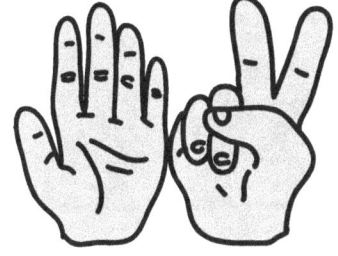 Seven 7

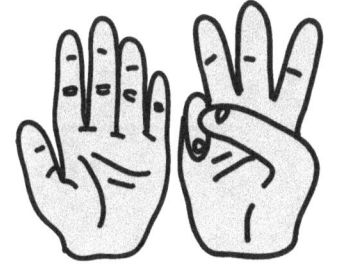 Eight 8

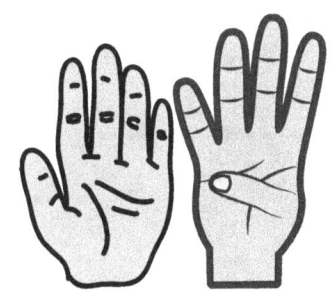 Nine 9

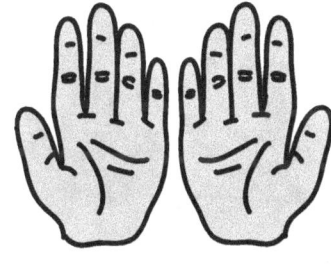 Ten 10

One

Retype the number

Math book for preschoolers 3-6 years old.

The book contains a group of mathematical activities to learn the basics of mathematics.

We have compiled a wonderful collection of fun and stimulating sports and educational activities for children.

Practical papers to learn writing numbers from 1 to 10.

Learn to read and write numbers from 1 to 100.

A group of activities to test the intelligence of your son and the math.

Helpful educational activities for math coloring, linking...

A simple way to learn additional operations.

We will also present the second and third levels of this book. They include mathematical operations, addition, subtraction, multiplication, and division.

By buguettaya ahlem

Specialized Mathematics and computer science

 One 1

 Two 2

 Three 3

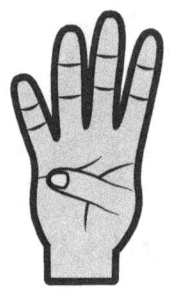 Four 4

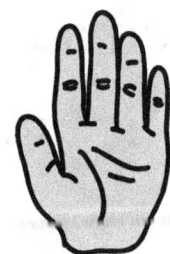 Five 5

Retype the number

Three

Retype the number

Calculate and then type the appropriate number within the frame

Calculate and then type the appropriate number within the frame

Calculate and link to the appropriate number

1

3

2

Complete

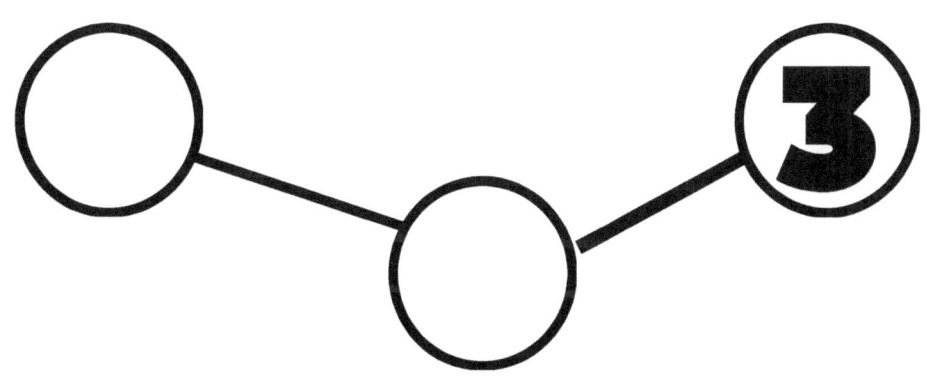

Four

Retype the number

Retype the number

Retype the number

Calculate and then type the appropriate number within the frame

Calculate and then type the appropriate number within the frame

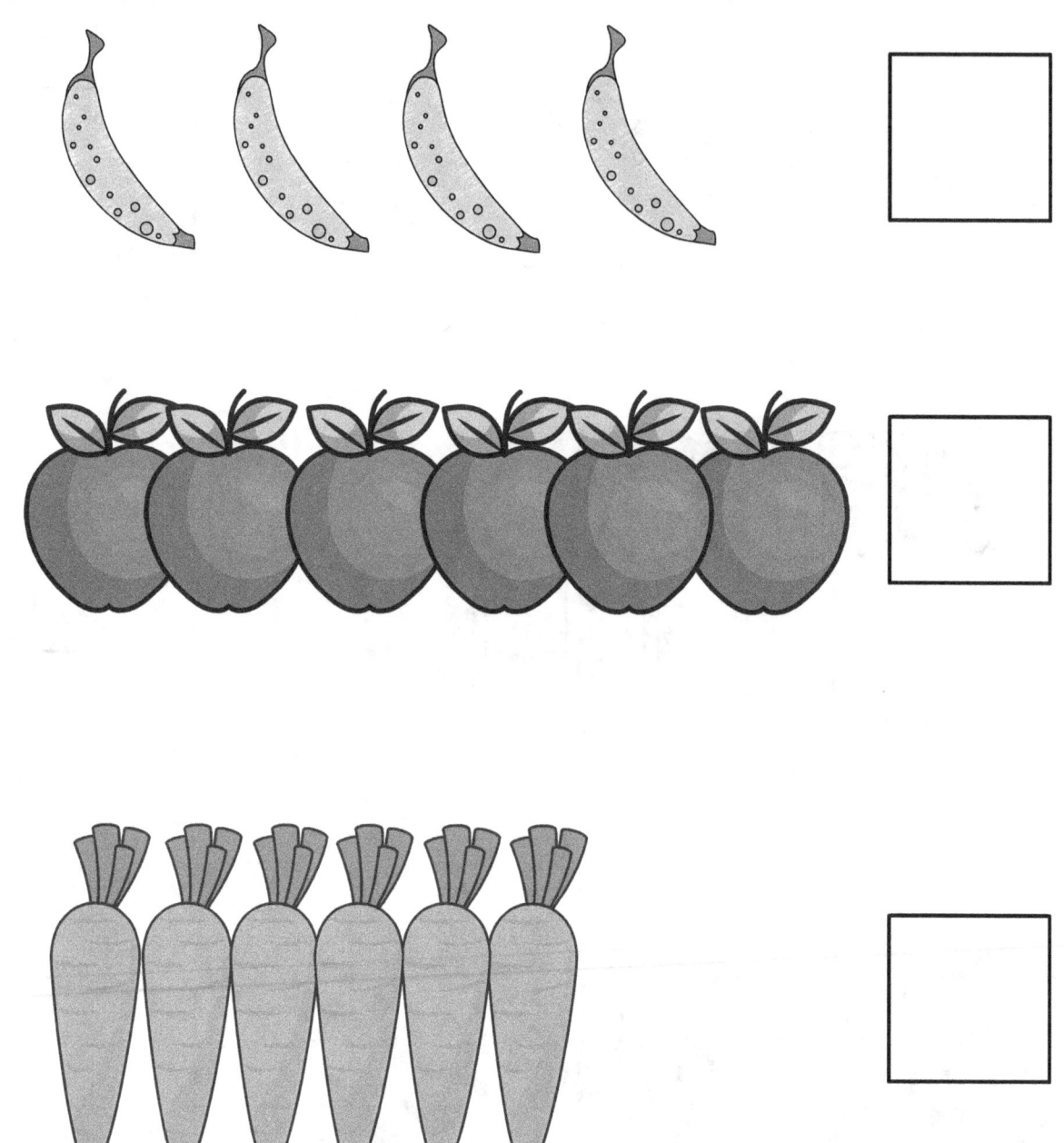

Calculate and link to the appropriate number

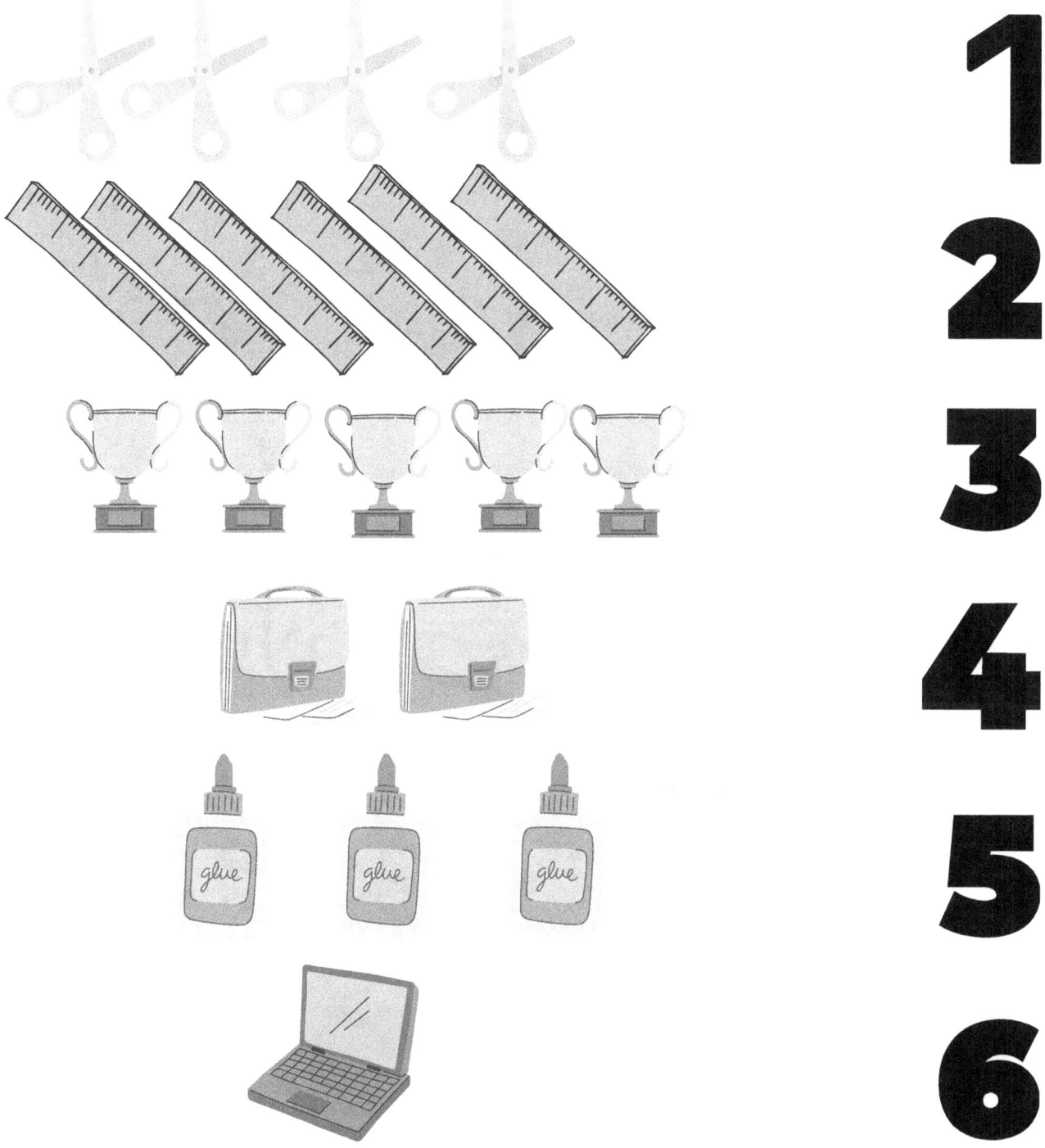

Complete

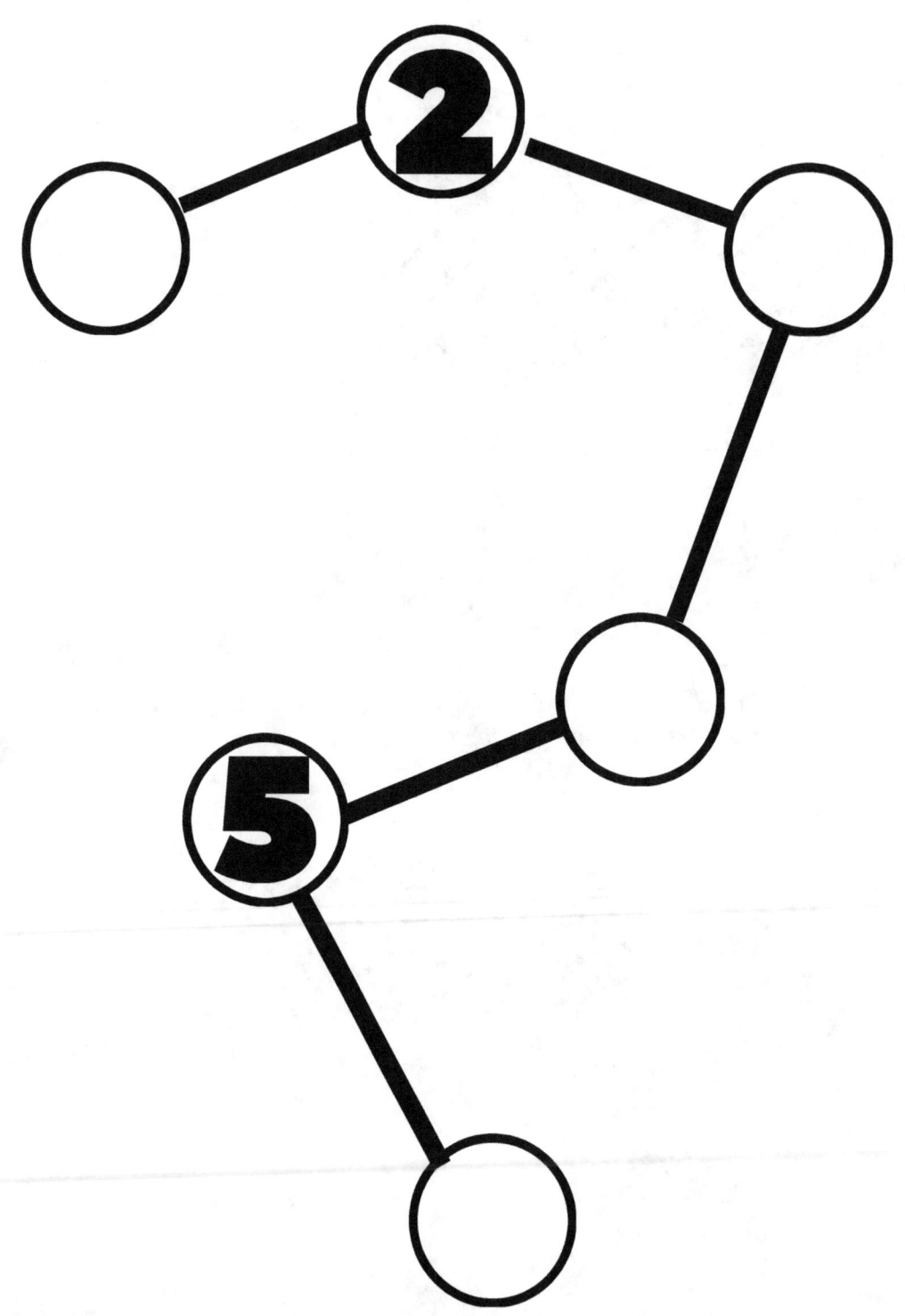

Seven

Retype the number

Eight

Retype the number

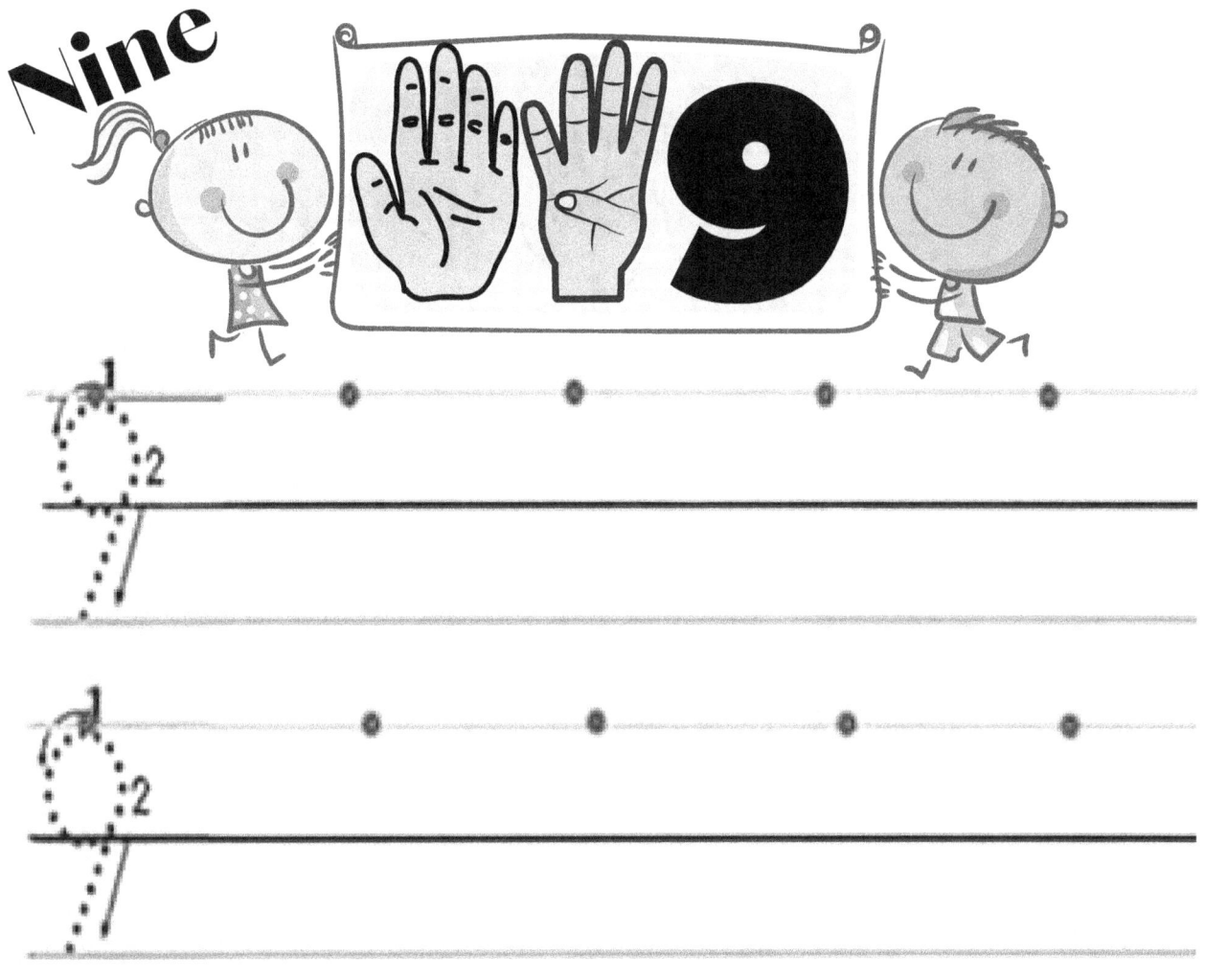

Retype the number

Ten

Retype the number

Calculate and then type the appropriate number within the frame

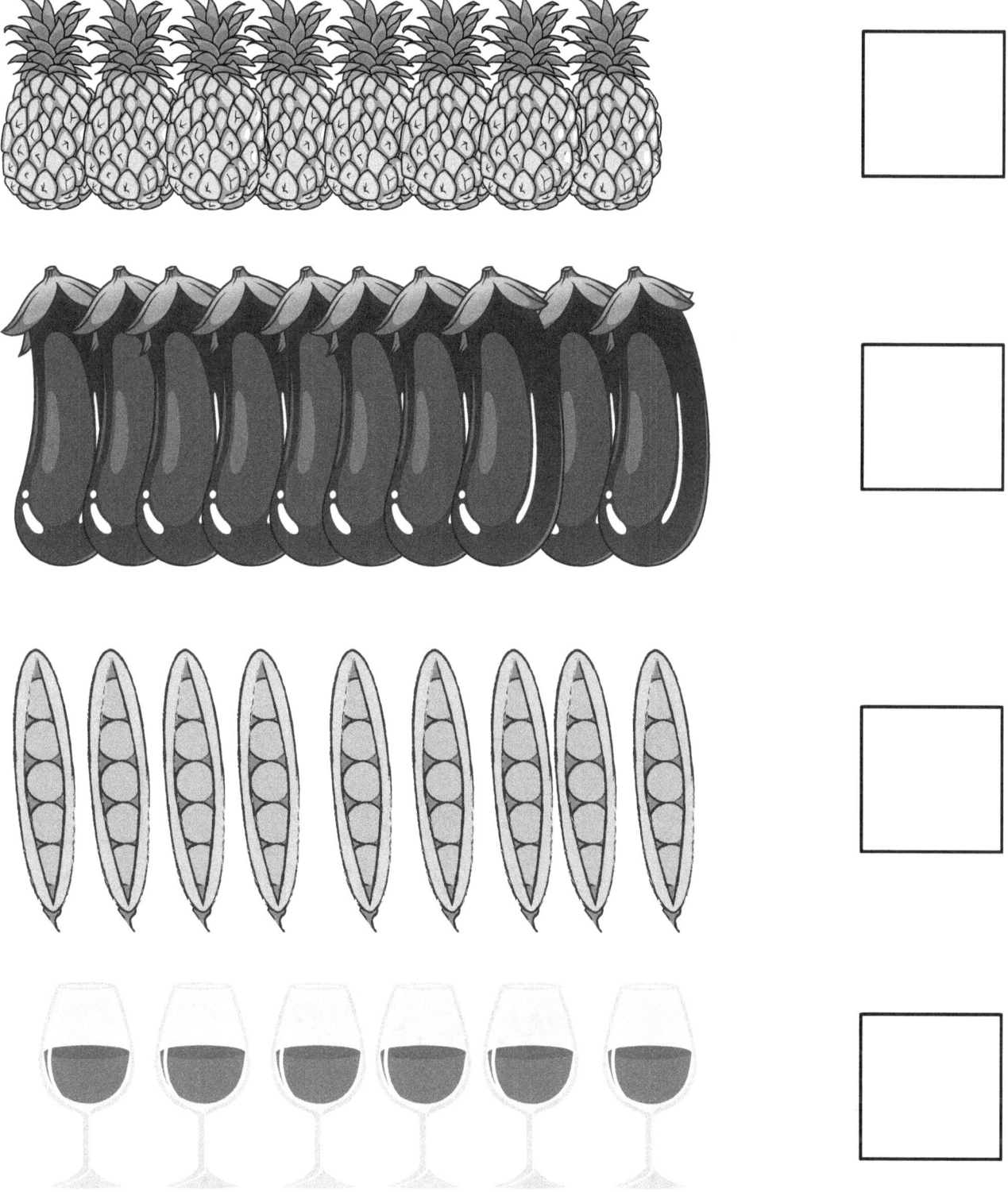

Calculate and then type the appropriate number within the frame

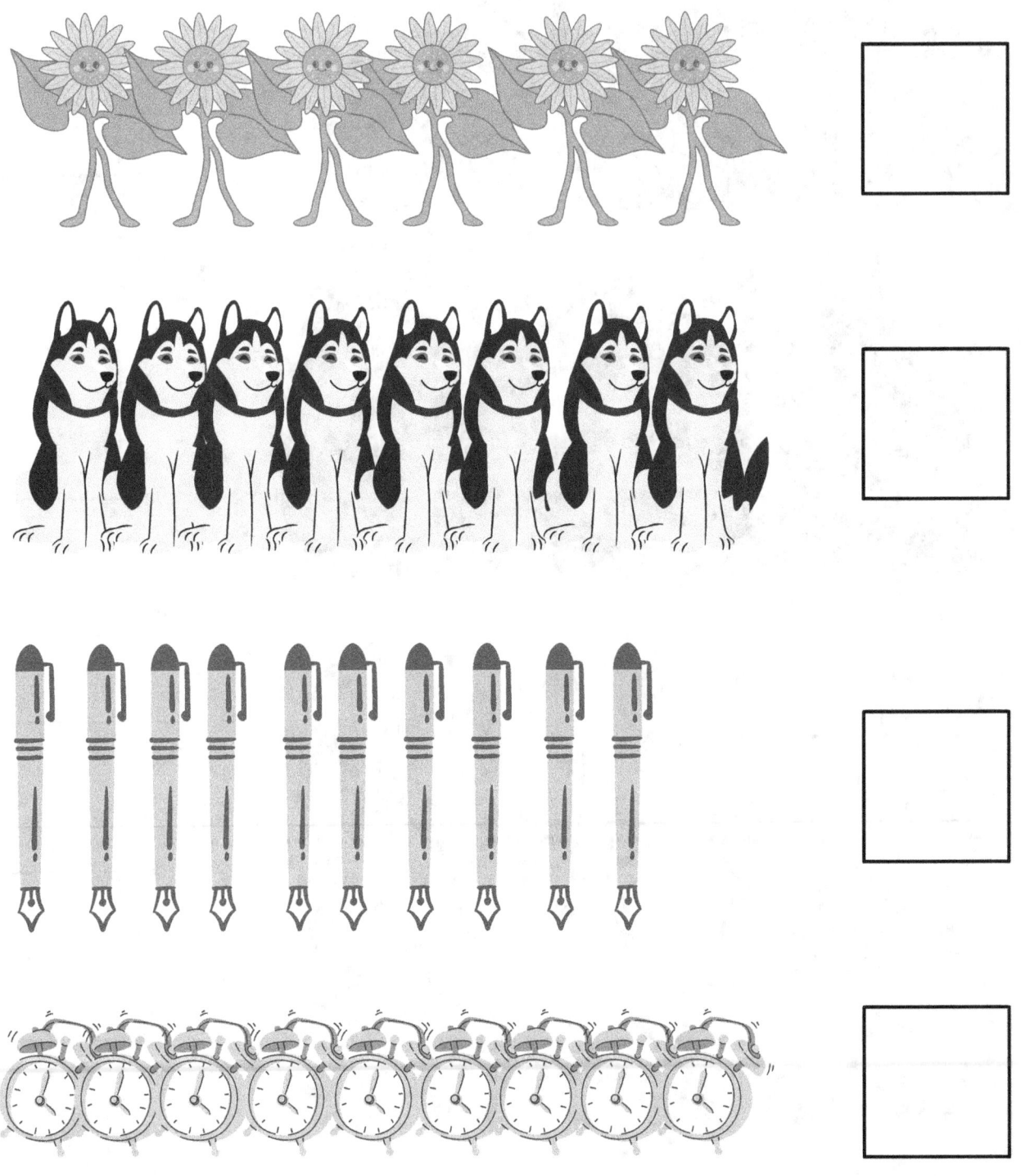

Calculate and link to the appropriate number

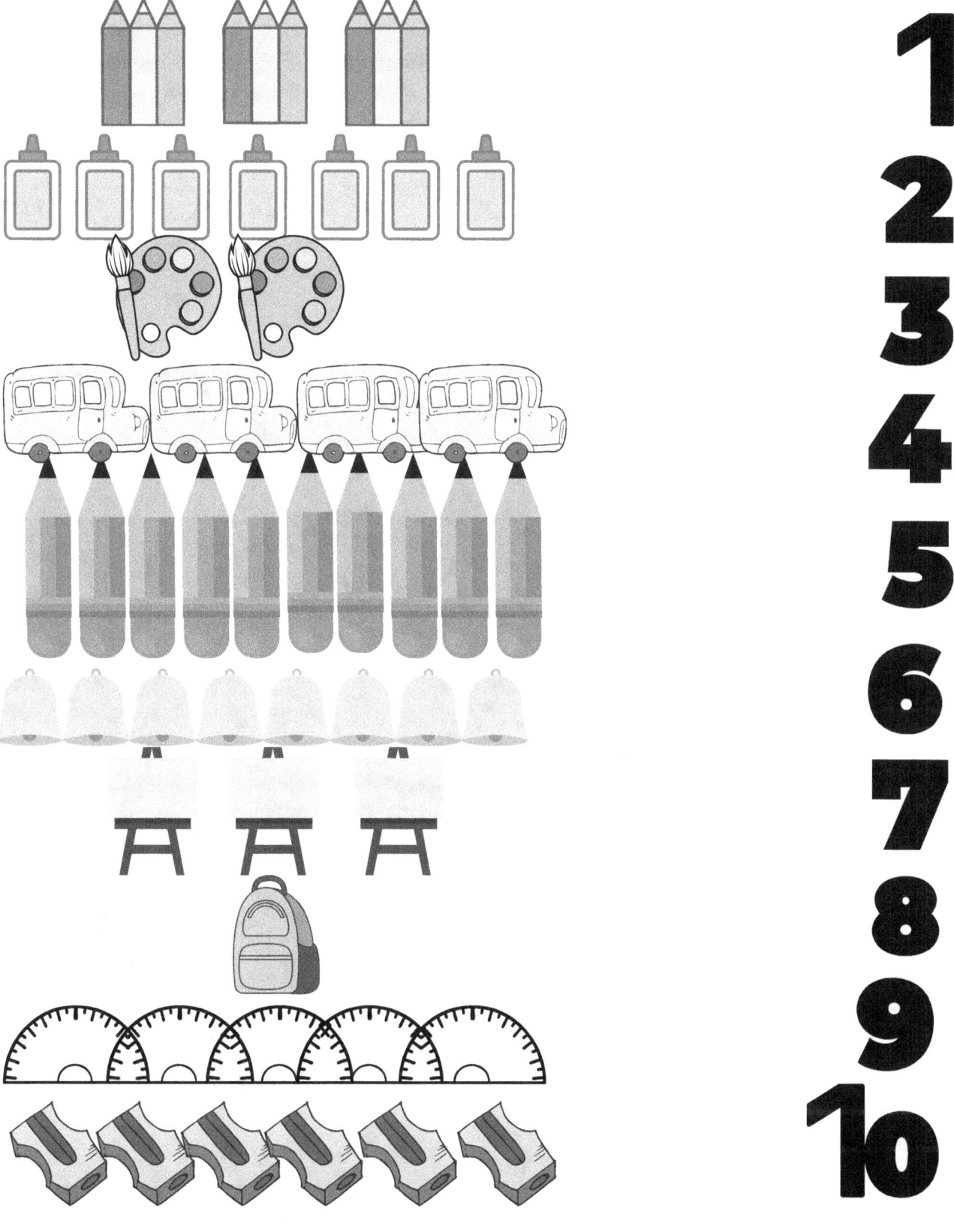

Complete

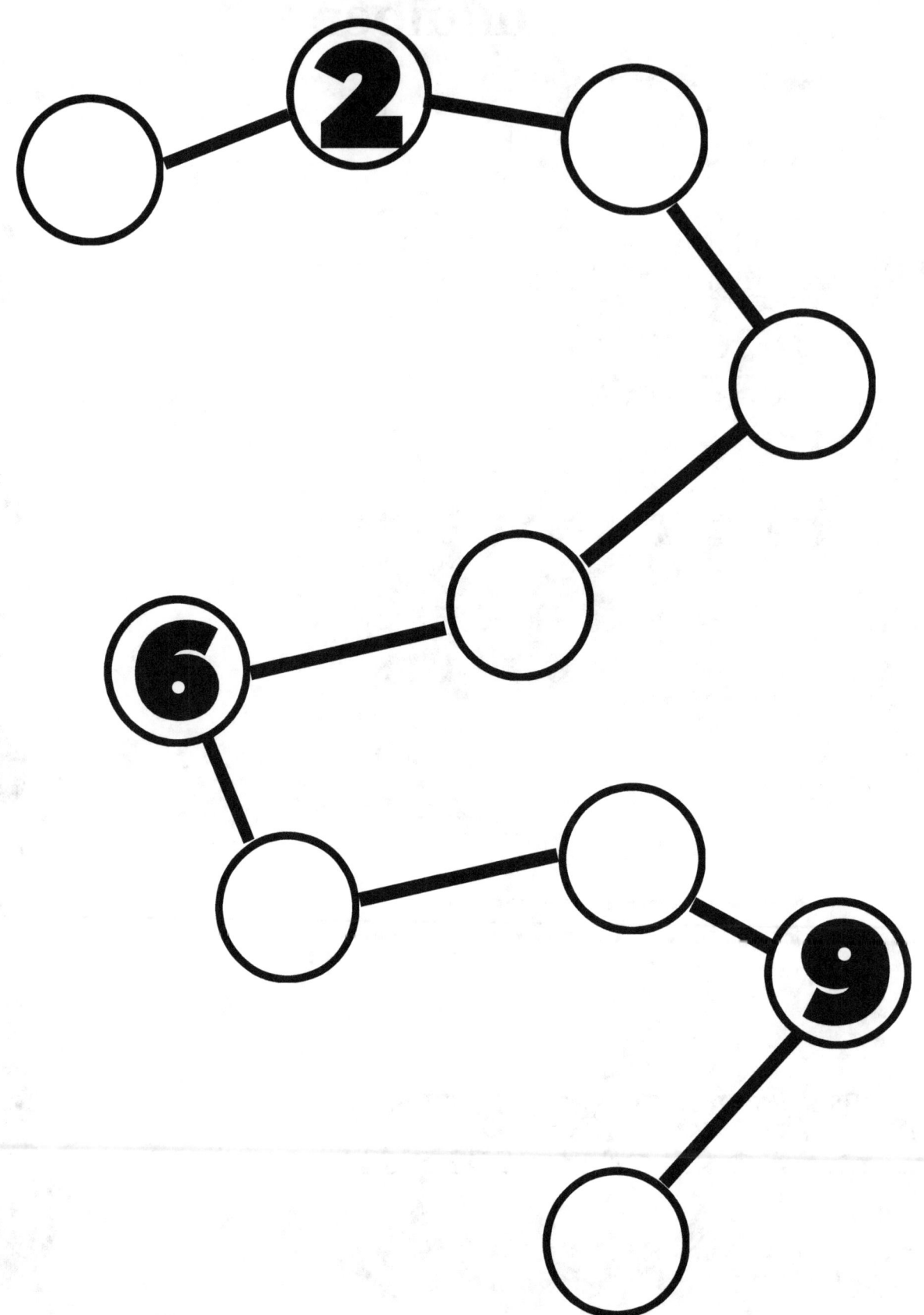

Activities

LEARN ARITHMETIC

Calculate the number of objects in the space and write the appropriate number

Calculate the number of objects in the space and write the appropriate number

Calculate the number of objects in the space and write the appropriate number

Complete

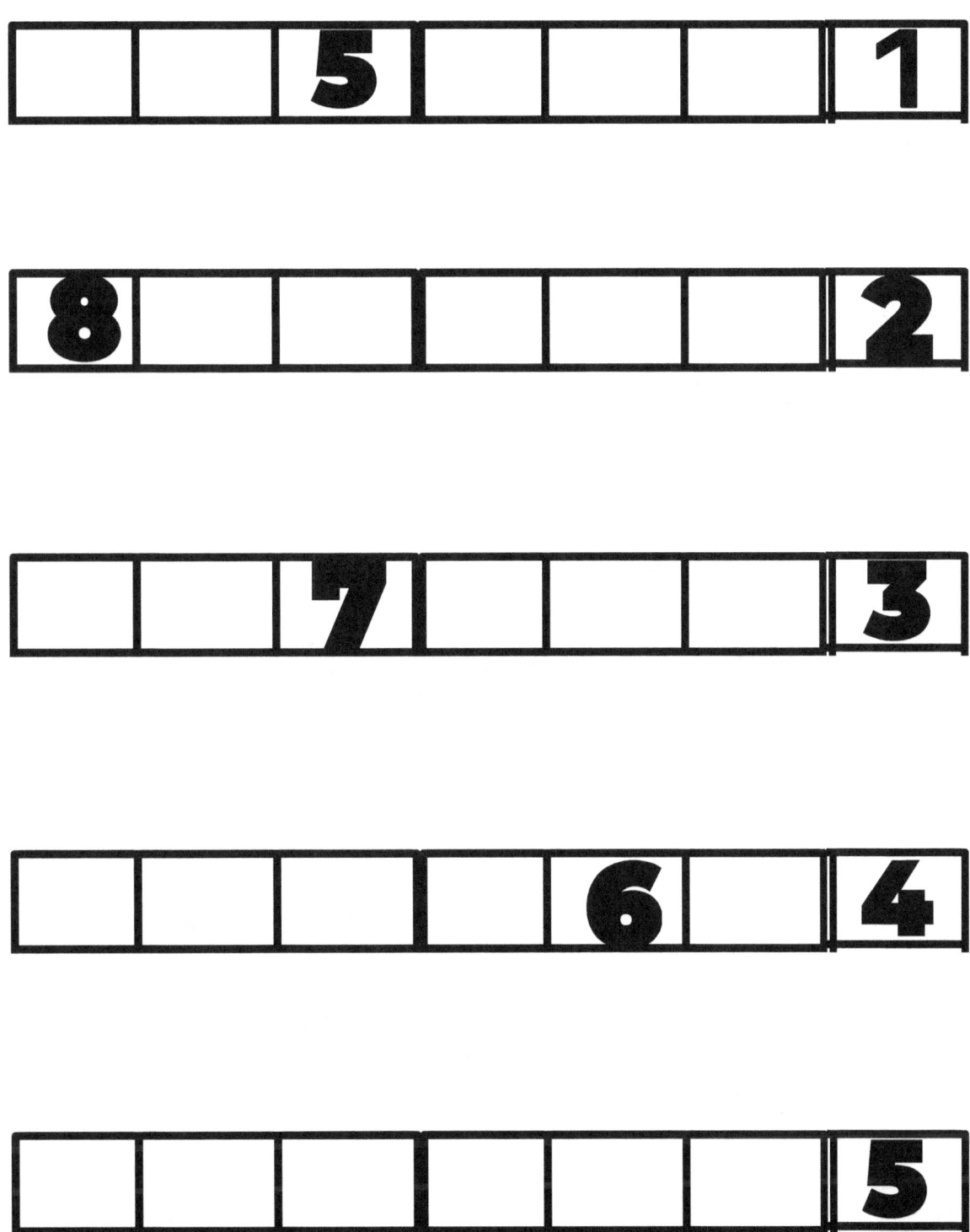

Put all 6 cats in a compartment

Put every two dogs in a compartment

Color the drawing by following the following color guide

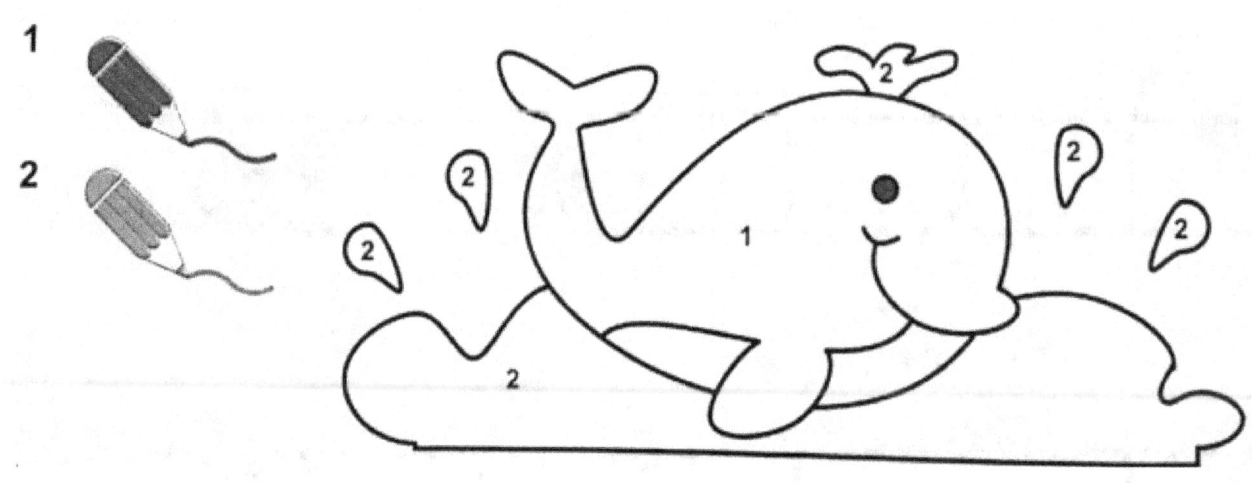

Put all 4 horses into a compartment

Put all 7 chickens into a compartment

Put all 5 computers in a compartment

Put all 3 wallets into a compartment

Help the bee by following the number 1 to get to the honey

2		2	10	2	6	1
10	1	1	1		9	1
3	1	5	1	5	1	1
	1	6	1		1	2

Connect the dots following the numbers from 1 to 9 to complete the form

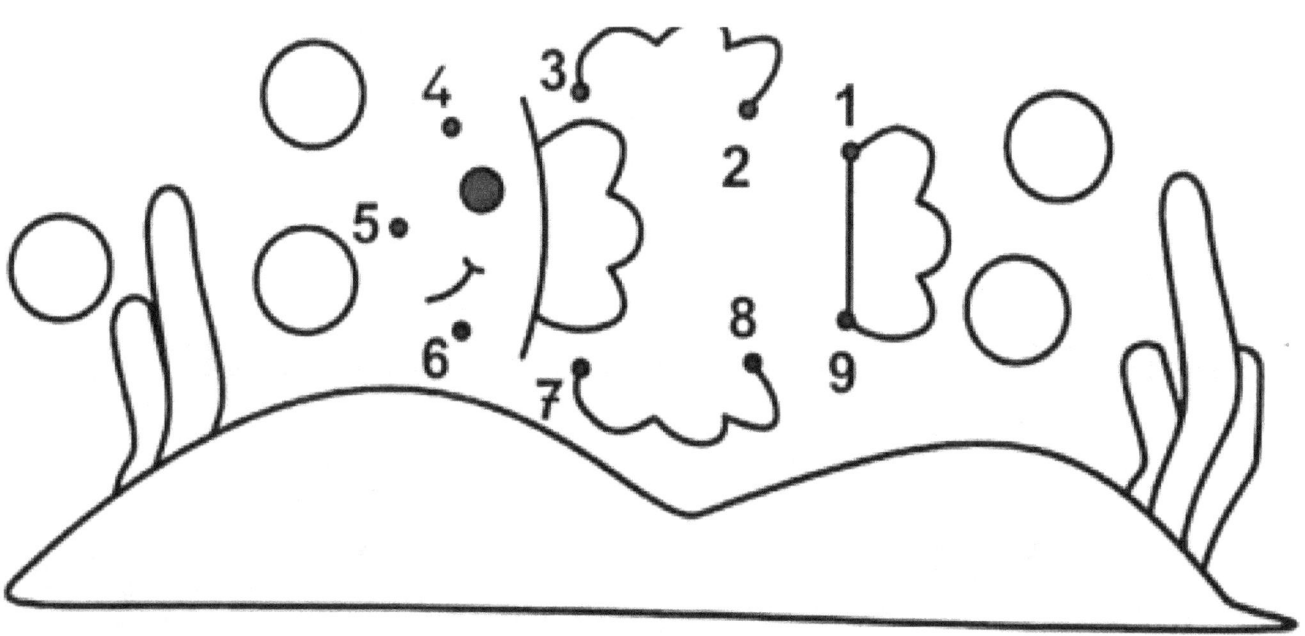

Color one chicken
Color 5 whales

Help the wolf by following the number
10 to get to the chicken

2		10	10	2	3	10
10	10	3	10		10	10
10	1	5	10	10	10	8
	5	6	3		1	9

Color the drawing by following the following color guide

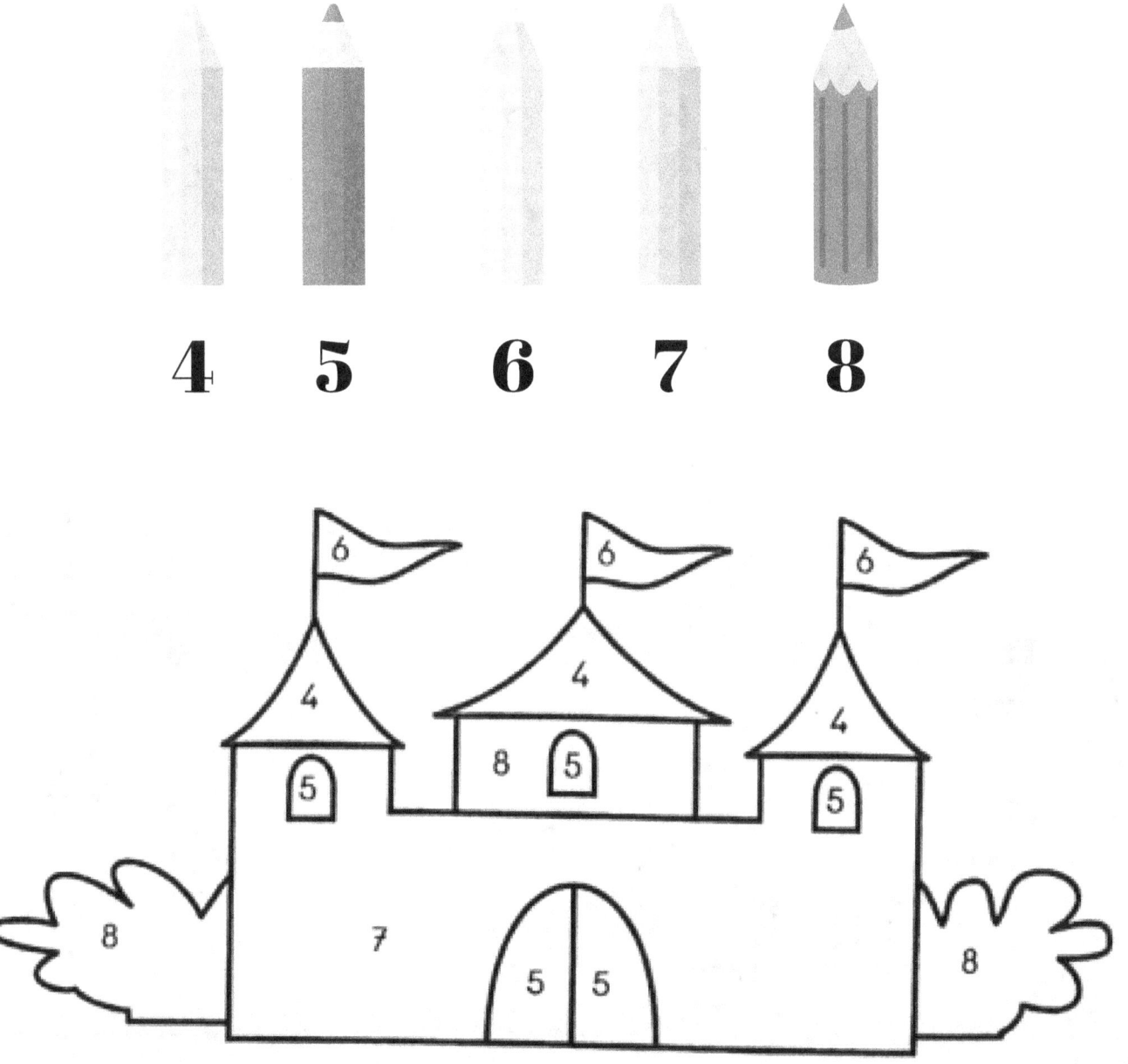

Color 8 butterflies

Help the bunny by following the number 4 to get to the carrots

6		2	1	5	4	4
4	4	4	2		4	7
4	9	4	9	4	3	2
	3	4	4	4	1	8

Color 9 camels

Help the cat by following the number 7 to get to the meat

2		9	1	10	7 ← 7
8	5	3	2		7	2
5	7	7	7	8	7	8
←		7	6	7		7	9

Learning numbers from 1 to 100

INTRODUCTION TO THE COLLECTION PROCESS

$10+1=11$ eleven

$10+2=12$ twelve

$10+3=13$ thirteen

$10+4=14$ fourteen

$10+5=15$ fifteen

$10+6=16$ sixteen

$10+7=17$ seventeen

$10+8=18$ eighteen

$10+9=19$ nineteen

$10+10=\boxed{20}$ twenty

20
=
10+10

$$20+1=21 \quad \text{twenty-one}$$

$$20+2=22 \quad \text{twenty-two}$$

$$20+3=23 \quad \text{twenty-three}$$

$$20+4=24 \quad \text{twenty-four}$$

$$20+5=25 \quad \text{twenty-five}$$

$20+6=26$ twenty-six

$20+7=27$ twenty-seven

$20+8=28$ twenty-eight

$20+9=29$ twenty-nine

$20+10=\boxed{30}$ thirty

$30+1=31$ thirty-one

$30+2=32$ thirty-two

$30+3=33$ thirty-three

$30+4=34$ thirty-four

$30+5=35$ thirty-five

$30+6=36$ **thirty-six**

$30+7=37$ **thirty-seven**

$30+8=38$ **thirty-eight**

$30+9=39$ **thirty-nine**

$30+10=\boxed{40}$ **forty**

$$40+1=41 \quad \text{forty-one}$$

$$40+2=42 \quad \text{forty-two}$$

$$40+3=43 \quad \text{forty-three}$$

$$40+4=44 \quad \text{forty-four}$$

$$40+5=45 \quad \text{forty-five}$$

$40+6=46$ forty-six

$40+7=47$ forty-seven

$40+8=48$ forty-eight

$40+9=49$ forty-nine

$40+10=\boxed{50}$ fifty

$50+1=51$ **fifty-one**

$50+2=52$ **fifty-two**

$50+3=53$ **fifty-three**

$50+4=54$ **fifty-four**

$50+5=55$ **fifty-five**

$$50+6=56 \quad \text{fifty-six}$$

$$50+7=57 \quad \text{fifty-seven}$$

$$50+8=58 \quad \text{fifty-eight}$$

$$50+9=59 \quad \text{fifty-nine}$$

$$50+10=\boxed{60} \quad \text{sixty}$$

$60+1=61$ sixty-one

$60+2=62$ sixty-two

$60+3=63$ sixty-three

$60+4=64$ sixty-four

$60+5=65$ sixty-five

$60+6=66$ sixty-six

$60+7=67$ sixty-seven

$60+8=68$ sixty-eight

$60+9=69$ sixty-nine

$60+10=\boxed{70}$ seventy

70+1=71 seventy-one

70+2=72 seventy-two

70+3=73 seventy-three

70+4=74 seventy-four

70+5=75 seventy-five

$70 + 6 = 76$ seventy-six

$70 + 7 = 77$ seventy-seven

$70 + 8 = 78$ seventy-eight

$70 + 9 = 79$ seventy-nine

$70 + 10 = \boxed{80}$ eighty

$$80+1=81 \quad \text{eighty-one}$$

$$80+2=82 \quad \text{eighty-two}$$

$$80+3=83 \quad \text{eighty-three}$$

$$80+4=84 \quad \text{eighty-four}$$

$$80+5=85 \quad \text{eighty-five}$$

$$80 + 6 = 86 \quad \text{eighty-six}$$

$$80 + 7 = 87 \quad \text{eighty-seven}$$

$$80 + 8 = 88 \quad \text{eighty-eight}$$

$$80 + 9 = 89 \quad \text{eighty-nine}$$

$$80 + 10 = \boxed{90} \quad \text{ninety}$$

90
=

$$10+10+10$$
$$+10+10+10$$
$$+10+10+10$$

$90+1=91$ ninety-one

$90+2=92$ ninety-two

$90+3=93$ ninety-three

$90+4=94$ ninety-four

$90+5=95$ ninety-five

$90+6=96$ ninety-six

$90+7=97$ ninety-seven

$90+8=98$ ninety-eight

$90+9=99$ ninety-nine

$90+10=100$ one hundred